SO-BDN-485

Is God Unnecessary?

Why Stephen Hawking Is Wrong
according to the Laws of Physics

WALTER ALAN RAY

iUniverse, Inc.
Bloomington

Is God Unnecessary?
Why Stephen Hawking Is Wrong according to the Laws of Physics

Copyright © 2012 Walter Alan Ray

All rights reserved. No part of this book may be used or reproduced by
any means, graphic, electronic, or mechanical, including photocopying,
recording, taping or by any information storage retrieval system
without the written permission of the publisher except in the case
of brief quotations embodied in critical articles and reviews.

iUniverse books may be ordered through booksellers or by contacting:

iUniverse
1663 Liberty Drive
Bloomington, IN 47403
www.iuniverse.com
1-800-Authors (1-800-288-4677)

Because of the dynamic nature of the Internet, any Web addresses or
links contained in this book may have changed since publication and
may no longer be valid. The views expressed in this work are solely those
of the author and do not necessarily reflect the views of the publisher,
and the publisher hereby disclaims any responsibility for them.

Any people depicted in stock imagery provided by Thinkstock are models,
and such images are being used for illustrative purposes only.

Certain stock imagery © Thinkstock.

ISBN: 978-1-4759-5463-0 (sc)
ISBN: 978-1-4759-5464-7 (e)

Library of Congress Control Number: 2012918639

Printed in the United States of America

iUniverse rev. date: 10/10/2012

Contents

Preface

Man is equally incapable of seeing the nothingness from
which he emerges and the infinity in which he is engulfed.

—Blaise Pascal, *Pensées*

I was walking across the Charles River Bridge, which connects Boston
to Cambridge. It was about one in the morning, and I was heading
back to MIT, where I was a sophomore in electrical engineering. It
was a beautiful, clear, starlit night. Looking up at the stars, a thought
entered my mind: "Walter, what if there is a God?" The question was
to myself because I considered myself an agnostic. My reply was this:
"No, I do not think there is, because he has never tried to communicate
with me." I arrived back at my fraternity house and did not think about
God much for the next several years.

I have written an Appendix entitled "The God Experiment" to
comment upon my journey from agnosticism.

One of the truly stunning statements I have heard recently in
the world of science came from Stephen Hawking, the most famous

scientist living in the twenty-first century. He stirred up a great amount of discussion and media response with his latest book, *The Grand Design*.[1] In this book Hawking says the laws of physics show us that God is unnecessary because the laws of physics explain the origin of our universe.

I decided to get *The Grand Design* and read about this remarkable statement. I read the book and afterward did not think Hawking's statement was supportable by the laws of physics that he selected. I read the book again carefully and decided I wanted to pursue Hawking's thesis about the laws of physics showing that God is unnecessary.

What could possibly have equipped me to write a book showing the greatest living scientist in the world to be mistaken in saying that the laws of physics show God is unnecessary? Here is a partial answer to this question.

I have long been interested in science, learning, and teaching. While I was in graduate school at MIT, I had the opportunity as a teaching assistant to teach MIT sophomores the basic course in electrical engineering.

After earning bachelor's and master's degrees from MIT, I worked at Microdot Inc. for several years as an electrical engineer and as a project engineer. While working at Microdot I designed a magnetron high-power generator that doubled the power output of the device that Microdot was using at that time.

Following this, I tapped into another interest that led to my earning a master of divinity degree from Fuller Theological Seminary in Pasadena, California, and a doctoral degree in biblical studies from

---

1   Stephen Hawking and Leonard Mlodinow, *The Grand Design* (New York: Bantam Books, 2010). In the interest of efficiency, when quoting from this book hereafter, I use the words "Hawking says" rather than cite the double authorship each time.

Princeton Theological Seminary in Princeton, New Jersey. While I was in graduate school at Princeton Theological Seminary, I had the privilege of teaching Greek to first-year students. Teaching electrical engineering and Greek each involved doing further research and study in preparing myself to teach, and each was an experience I thoroughly enjoyed.

Then I served as pastor of Glenkirk Presbyterian Church in Glendora, California.

During all this time, I held on to a deep interest in our cosmos, and since my years in college, I have enjoyed studying and reading about the works of our leading cosmologists and physicists.

This brings us back to Hawking and the laws of physics.

Stephen Hawking is so well known that when he spoke at the California Institute of Technology in January 2011, people waited in line for two hours to get in before the talk began. He was the rock star of the physics world. Because he has received so much well-deserved fame from his work, many nonscientists take his pronouncements as if they are scientific fact and assume he is correct when he says that the laws of physics make it unnecessary to postulate the existence of God.

For this reason, the idea of presenting the other side of Hawking's arguments in *The Grand Design* germinated in my mind, and I carefully read his book a third time. I decided it was possible for me to disprove Hawking's statement about the laws of physics showing that God is not necessary. I have attempted to do this not by using philosophical or theological arguments but strictly by using the same laws of physics that Hawking says demonstrate his position.

Introduction

It is the purpose of *Is God Unnecessary?* to examine the laws of physics that Hawking sets forth in his book *The Grand Design*.

In chapter 1 we look at Hawking's *Apparent Miracle*. Hawking selected an unusual title for chapter 6 of his book *The Grand Design*. He called it "The Apparent Miracle." By this phrase, Hawking refers to all the seeming coincidences that are necessary for life to be able to exist upon earth. These coincidences are also referred to as the *fine-tuning* of our universe, or the *Goldilocks effect* (because everything is just right). Hawking says:

> What can we make of these coincidences? Luck in the precise form and nature of fundamental physical law is a different kind of luck from the luck we find in environmental factors. It cannot be so easily explained, and has far deeper physical and philosophical implications. Our universe and its laws appear to have a design that is tailor-made to support us and, if we are to exist, leave little room for alteration. That is not easily explained, and raises the natural question of why it is that way.[2]

---

2   Hawking, *The Grand Design*, 162.

Hawking says that the existence of our universe can be fully accounted for by the laws of physics.

> M-theory predicts that a great many universes were created out of nothing. Their creation does not require the intervention of some supernatural being or god. Rather, these multiple universes arise naturally from physical law.[3]

There are three major questions about Hawking's proposal: How do we explain the Apparent Miracle? How did human life begin? Can something come from nothing? Chapters 1, 4, and 5 look at some of the factors involved in the Apparent Miracle. Chapter 2 looks at the origin of human life, and chapter 3 looks at whether or not something can come from nothing.

In chapter 2 we examine Hawking's assumption that Charles Darwin explained the origin of human life and that this should encourage us to believe that in an analogous fashion science can come up with an understanding of the origin of the universe. Hawking says:

> Just as Darwin and Wallace explained how the apparently miraculous design of living forms could appear without intervention by a supreme being, the multiverse concept can explain the fine-tuning of physical law without the need for a benevolent creator who made the universe for our benefit.[4]

Is Hawking's understanding of Darwin accurate? Does Hawking's understanding of Darwin give added support to the idea that we can explain the origin of the universe by the laws of physics? Did Charles

---

3   Hawking, *The Grand Design*, 8–9.
4   Hawking, *The Grand Design*, 165.

Darwin have a theory about the origin of life? I believe that Darwin's theory of evolution is an established scientific theory, and I accept Darwin's own understanding of his theory of evolution.

In chapter 3 we examine the question "Can something come out of nothing?" Hawking says the answer is yes. He says that quantum fluctuations lead to the creation of the universe from nothing, and that quantum fluctuations are caused by virtual particles that exist in a vacuum. Is this idea consistent with the laws of physics? Hawking proposes what he calls a *no-boundary condition* of our universe. This would eliminate the perplexing question: How did our universe begin? Are these two theories supported and established by the laws of physics?

Chapter 4 looks at the factor that Hawking considers the most impressive coincidence of all, the cosmological constant in Einstein's equations, which determines whether our universe is static, contracting, or expanding. How is it that the cosmological constant turns out to have a value that is incredibly small, and *must* be this small in order for our universe to be such that life can begin, evolve, and result in human beings? The cosmological constant is one part of the Apparent Miracle. Many physicists and cosmologists today consider this long-standing question the most difficult as-yet-unanswered question in physics. Leonard Susskind says:

> The modern principles of physics are based on two foundations: the Theory of Relativity and quantum mechanics. The generic result of a world based on these principles is a universe that would very quickly self-destruct. But for reasons that have been completely incomprehensible, the cosmological constant has been fine-tuned to an astonishing degree. This, more than any

other "lucky" accident, leads some people to conclude that the universe must be the result of a design.[5]

In chapter 5 we discuss Hawking's solution to the "completely incomprehensible" value of the cosmological constant. Hawking believes that the number of different universes according to M-theory is about 10^{500}, and that from this enormous number there will surely be at least one universe that will have the right parameters for life. We will see how this relates to the work of Sir Roger Penrose concerning the odds of the Apparent Miracle coming about by chance.

Chapter 6 asks this question: Would an infinite number of universes be enough to guarantee that one of these universes would have a planet with the characteristics that could sustain life? Would an infinite number of universes solve the conundrums of Hawking's Apparent Miracle? What are some of the factors involved in the theory of an infinite number of universes?

Chapter 7 discusses how physics and mathematics join in showing that in the current state of our knowledge, physics and mathematics do have something to say about the origin of our universe. They narrow the possibilities down to two. This is a most surprising result, which I did not expect when I started this book.

The following list of approximate mathematical values of certain quantities gives the reader a feeling for some of the numbers mentioned in this book. In discussing very large or very small numbers I have used the standard powers of ten notation. The number 10^{10} means one followed by ten zeros, while 10^{-10} means $1/10^{10}$.

---

5   Leonard Susskind, *The Cosmic Landscape* (New York: Little, Brown and Company, 2006), 12. Leonard Susskind is a professor of theoretical physics at Stanford University, and is often considered the father of string theory.

Planck unit of time	10^{-43} sec
Size of universe at big bang	smaller than 10^{-33} cm
Planck unit of distance	10^{-33} cm
Mass of an electron	10^{-30} kg
Size of an atomic nucleus	10^{-13} cm
US national debt in dollars	10^{13}
Number of stars in our universe	10^{21}
Approximate odds against a wooden chair suddenly jumping one cm upward[6]	10^{60}
Number of particles in our universe	10^{80}
Number of universes according to M-theory	10^{500}

---

6   In a course at MIT, our physics professor actually calculated the approximate odds that at any given moment in time a chair would jump one centimeter off the floor without any outside influences. This was done by calculating the odds that the random motion of enough particles inside the chair could be in alignment with each other for a short moment in time to the degree that they could affect the motion of the solid itself.

CHAPTER 1

Hawking's Apparent Miracle

HAWKING BELIEVES THAT ACCORDING to M-theory, many universes were created out of nothing. Hawking says:

> M-theory predicts that a great many universes were created out of nothing. Their creation does not require the intervention of some supernatural being or god. Rather, these multiple universes arise naturally from physical law.[7]

According to Hawking, the natural laws of physics explain the origin of our universe and the origin of life. The earth is amazingly *fine-tuned* to support life. How is it that our planet earth came to meet the highly improbable combination of initial parameters that are necessary for life to exist on our planet earth? Hawking is so impressed by the fine-tuning of our universe that he calls it the *Apparent Miracle*. Hawking makes it clear in this book that he does not believe it is a miracle. But his use of the word "miracle" is significant in that he recognizes the challenge of explaining the fine-tuning by logic and the laws of physics.

---

7   Hawking, *The Grand Design*, 8–9.

Hawking's apparent miracle has also been called the *Goldilocks Enigma*,[8] summarized by the question "How did earth come to have precisely the necessary characteristics for supporting life?" How did all the factors necessary for life to exist come to be "just right?" How do we explain these highly improbable coincidences by the laws of physics?

A simple example of fine-tuning is the existence of water, and the present temperature range on the surface of the earth. Humans need water in order to survive. In order for water to exist in liquid form, the earth must be within a certain range of distances from the sun. There is a fine band of temperatures within which water can exist in liquid form: zero degrees centigrade to 100 degrees centigrade. Below this range we have ice, and above it we have steam, both of which are inconducive to life. The temperature range within which human life can exist (think back to the time before we had heaters!) is even narrower than that for water. The two planets adjacent to Earth are Venus and Mars. Venus has a high temperature of 860 degrees Fahrenheit. Martian surface temperatures vary from lows of about –125 degrees Fahrenheit during the polar winters (again, too cold for humans before we had heaters) to highs of up to 23 degrees Fahrenheit in summers.

Another example of fine-tuning is the existence of carbon. Scientists agree that chemically speaking, carbon is one of the ingredients necessary for life to exist. Up until 1951 scientists were unable to explain how carbon could have come to exist. It was in the early 1950s that Fred Hoyle, then an unknown British astronomer, came up with a theory

---

8   Paul Davies, *The Goldilocks Enigma* (Great Britain: Penguin Press, 2006). Davies is an internationally acclaimed cosmologist and physicist. Several examples of fine-tuning are given in *The Goldilocks Enigma* on pages 139–150. In this book he sets forth many of the ways that our planet earth meets the highly improbable conditions that are necessary for human life to exist.

that was later proven to be right. George Gamow, a famous Russian scientist, composed a tongue in cheek account of Hoyle's discovery:

> And so, with the help of God, Hoyle made heavy elements in this way, but it was so complicated that nowadays neither Hoyle, nor God, nor anyone else can figure out exactly how it was done.[9]

Various examples of fine-tuning are given by Hawking in his chapter titled "The Apparent Miracle." Hawking says:

> The laws of nature form a system that is extremely fine-tuned, and very little in physical law can be altered without destroying the possibility of the development of life as we know it. Were it not for a series of startling coincidences in the precise details of physical law, it seems, humans and similar life-forms would never have come into being.[10]

Leonard Susskind talks about how we cannot attribute the earth's ability to sustain life to lucky accidents. Susskind says:

> Our own universe is an extraordinary place that appears to be fantastically well designed for our own existence. This specialness is not something that we can attribute to lucky accidents, which is far too unlikely. The apparent coincidences cry out for an explanation.[11]

---

9   Davies, *The Goldilocks Enigma*, 139. There is an excellent discussion of what it is about the origin of carbon that was so difficult to understand in Davies. *Ibid*, 135–139.

10   Hawking, *The Grand Design*, p. 161.

11   Susskind, *The Cosmic Landscape*, p. 343.

At Least Thirty Knobs Fine-Tune Our Universe

Davies sums up the Goldilocks Enigma (which is Hawking's Apparent Miracle) by referring to the metaphor of a Designer Machine that has knobs that a Cosmic Designer can twirl to alter the various parameters of our universe, such as the masses of particles and the strengths of forces. Davies says:

> Returning to my Designer Machine metaphor, the collection of felicitous "coincidences" in physics and cosmology implies that the Great Designer had better set the knobs carefully, or the universe would be a very inhospitable place. How many knobs are there? The Standard Model of particle physics has about twenty undetermined parameters, while cosmology has about ten. All told, there are over thirty "knobs." ... some of the examples I have given demand "knob settings" that must be fine-tuned to an accuracy of less than 1 percent to make a universe fit for life.[12]

Is there any way we can quantify the Goldilocks Enigma? Can we come up with any numbers that give us the statistical odds for earth being conducive to life by chance? Yes, this can be done, as we shall see in chapters 4, 5, and 6. In chapter 2 we look at the second major question regarding Hawking's statement about the laws of physics: How did the origin of life occur?

---

12   Davies, *The Goldilocks Enigma*, 146.

CHAPTER 2

A Reference to Darwin by Hawking

STEPHEN HAWKING DRAWS A parallel between Darwin's theory of evolution and the origin of the universe, saying that Darwin showed how the origin of life could have occurred without any intervention by a supreme being. Hawking says:

> Just as Darwin and Wallace explained how the apparently miraculous design of living forms could appear without intervention by a supreme being, the multiverse concept can explain the fine-tuning of physical law without the need for a benevolent creator who made the universe for our benefit.[13]

If Darwin did indeed show that the origin of life could have taken place without a supernatural agency, then Hawking has a point in saying that the origin of the universe might also have taken place without any supernatural agency.

But there is a problem with Hawking's statement here. *It is a common misconception that Darwin's theory of evolution showed how the origin of life came about.* There is no place in *On the Origin of Species*

---

13   Hawking, *The Grand Design*, p. 165

where Darwin gives any kind of scientific theory about the *origin* of life. Richard Dawkins is one of the strongest supporters of the theory that human life began by itself, and yet Dawkins agrees that Darwin's famous work, *On the Origin of Species*, does not discuss how evolution began. Dawkins says:

> Darwin didn't discuss how evolution began in *On the Origin of Species*. He thought the problem was beyond the science of his day … Darwin went on to say, "It is mere rubbish, thinking at present of the origin of life."[14]

Richard Dawkins finds plausible the *RNA world hypothesis* of the origin of life. Andrew H. Knoll, a Professor at Harvard, discusses some of the problems with the RNA world hypothesis and concludes by saying:

> The problems are so difficult that many researchers have given up on the idea that RNA was the primordial molecule of life.[15]

Knoll then comments further on the problem of how life originated, and says:

> We are not close to solving the riddle of life's origins. Origin-of-life research resembles a maze with many entries, and we simply haven't traveled far enough down most routes to know which ones end in blind alleys … there is a direct route through the maze, if only we can find it.[16]

---

14   Richard Dawkins, *The Greatest Show on Earth* (New York: Free Press, 2009), 417.

15   Andrew H. Knoll, *Life on a Young Planet* (New Jersey: Princeton University Press, 2003), 75–79.

16   Knoll, *Life on a Young Planet,* 88.

Frances Collins is one of the US's leading geneticists. For many years he has directed the Human Genome Project, where he led the work of hundreds of scientists. Collins has written a book where he discusses the origin of life, wherein he is in complete agreement with Knoll. Collins says:

> How did self-replicating organisms arise in the first place? It is fair to say that at the present time we simply do not know. No current hypothesis comes close to explaining how in the space of a mere 150 million years, the prebiotic environments that existed on planet earth gave rise to life.[17]

Dawkins gives his own theory of how life on earth began by referring to statistics. Dawkins assumes the number of planets in our universe to be a billion billion. Then he says:

> Now, suppose the origin of life, the spontaneous arising of something equivalent to DNA, really was a quite staggeringly improbable event. Suppose it was so improbable as to occur on only one in a billion planets ... And yet, even with such absurdly long odds, life will still have arisen on a billion planets—of which Earth, of course, is one.[18]

Dawkins is arguing that in our universe, life will have arisen spontaneously on a billion planets. Mathematics and physics strongly suggest that Dawkins's argument is fallacious. Dawkins has not taken into account Hawking's Apparent Miracle. Perhaps Dawkins is unaware

---

17   Frances Collins, *The Language of God* (New York: Simon & Schuster, Inc., 2006), 90.

18   Richard Dawkins, *The God Delusion* (Great Britain: Bantam Press, 2006), 137–138.

of the current work of physicists on the cosmological constant. We will see in chapter 3 that Hawking shows that the odds against the chance occurrence of the cosmological constant of earth being such that life can be sustained are at least 10^{120} to one! This means that we would require at least 10^{120} planets in our universe in order to make it an even chance that one of these planets could sustain life. But we only have 10^{18} planets according to Dawkins. Therefore the odds *against* one of our 10^{18} planets spontaneously giving birth to life are about 10^{100} to one. This number is far too large to assume that life could occur by chance. As quoted above, both Hawking and Susskind say that a number on this order of magnitude requires some answer other than "it occurred by chance." It may be that someday science will come up with a scientific explanation for the origin of life. Someday the theory of evolution may show how nonlife transitioned into life. But that has not happened yet.

Hawking errs when he says Darwin showed how the apparently miraculous design of living forms could appear without intervention by a supreme being. Darwin did not show how living forms could appear without intervention by a supreme being, because Darwin had no theory at all for the origin of life.

Let us now turn to the second big question about Hawking's theory. Can something come from nothing?

CHAPTER 3

Can Something Come from Nothing?

HAWKING SAYS THAT SOMETHING can come out of nothing. In fact, he says that many somethings can come out of nothing. Hawking says:

> Quantum fluctuations lead to the creation of tiny universes out of nothing. A few of these reach a critical size then expand in an inflationary manner, forming galaxies, stars, and, in at least one case, beings like us.[19]

Using standard guidelines of English grammar and interpretation on this statement by Hawking, if quantum fluctuations lead to the creation of tiny universes, then quantum fluctuations existed *before* the creation of tiny universes took place. This would mean that quantum fluctuations existed before the big bang. Hawking says:

> The first actual scientific evidence that the universe had a beginning came in the 1920s ... In 1929 [Edwin Hubble] published a law relating [the galaxies'] rate of recession to their distance from us, and concluded that the universe is expanding. If that is true, then the universe must have been smaller in the past. In fact, if we extrapolate to the

---

19   Hawking, *The Grand Design*, 137.

distant past, all the matter and energy in the universe would have been concentrated in a very tiny region of unimaginable density and temperature, and if we go back far enough, there would a time when it all began – the event we now call the big bang.[20]

One could not say "quantum fluctuations lead to the creation of tiny universes" unless one believes that these quantum fluctuations exist before the creation of these tiny universes. If these tiny universes are created, then this means that there is a time when these tiny universes do not exist, but what creates them does exist! So these quantum fluctuations exist *before* the creation of the tiny universes.

Empty Space Is Not Really Empty

Hawking then explains the source of these quantum fluctuations. He says that *empty space* is not really empty. It is filled with virtual particles. Hawking says that these virtual particles must exist in empty space as a result of the Heisenberg uncertainty principle:

> The value of a field and its rate of change play the same role as the position and velocity of a particle. That is, the more accurately one is determined, the less accurately the other can be. An important consequence of that is that there is no such thing as empty space. That is because empty space means that both the value of a field and its rate of change are exactly zero. Since [Heisenberg's] uncertainty principle does not allow the values of both the field and the rate of change to be exact, space is never empty. It can have a state of minimum energy, called the vacuum, but that state is subject to what are called quantum jitters, or

20   Hawking, *The Grand Design*, 124.

vacuum fluctuations, particles and fields quivering in and out of existence.[21]

If the Heisenberg uncertainty principle makes it impossible for space to be completely empty, then at that earliest moment when space tried to be empty it could not be so because the Heisenberg uncertainty principle required that there be virtual particles. One question raised by this statement is this: How did it come to be that the Heisenberg principle already existed? Where did the Heisenberg uncertainty principle come from? We might someday have a scientific explanation for this, but we do not have one yet.

Hawking says that the presence of virtual particles in empty space has been proven by actual measurements of their effect on neighboring particles:

> One can think of the vacuum fluctuations as pairs of particles that appear together at some time, move apart, then come together and annihilate each other … These particles are called virtual particles. Unlike real particles, virtual particles cannot be observed directly with a particle detector. However, their indirect effects, such as small changes in the energy of electron orbits, can be measured, and agree with theoretical predictions to a remarkable degree of accuracy.[22]

Quantum fluctuations occur as a result of *quantum jitters*, when these elementary particles pop in and out of existence, each one lasting about a billion trillionth of a second. These short-lived particles fill the vacuum, and cause the vacuum to have energy. It is these virtual particles that somehow resulted in all the energy in our universe being

---

21 Hawking, *The Grand Design*, 113.
22 Hawking, *The Grand Design*, 113.

concentrated in a very small space. Out of this great amount of energy there came the big bang. So, according to Hawking, part of the reason that something can come out of nothing is that so-called empty space is not really empty, but is filled with virtual particles.

Hawking says that the effect of virtual particles on other particles has been measured. That is true,[23] but these measurements were made *after* the big bang occurred. Remember, Hawking said, "quantum fluctuations lead to the creation of tiny universes out of nothing." If quantum fluctuations lead to the creation of tiny universes out of nothing, then these quantum fluctuations must have existed before the big bang. Yes, we have measurements showing that virtual particles exist. But these measurements were made *after* the big bang. This does not mean that virtual particles existed *before* the big bang, and in fact caused the big bang. It simply means that virtual particles exist *after* the big bang. We do not know if any virtual particles existed before the big bang, and we have no way of knowing what, if anything, existed before the big bang. Let me explain the reason for this last statement.

Scientists are in general agreement that there is something unique about *Planck time*. Many scientists consider it to be the smallest possible unit of time. A unit of Planck time is defined as 10^{-43} seconds. The *Planck epoch* normally refers to the first 10^{-43} seconds after the big bang. Physicists presently believe that the laws of physics would break down during the Planck epoch. The highest possible temperature, known as the *Planck temperature*, is considered to be 10^{32} kelvin, and it occurred during the Planck epoch.[24] We do not know how anything behaves at this elevated temperature, but we know it is hot enough to melt anything. If there were any information carried by particles, or if

23   Brian Greene, *The Fabric Of The Cosmos* (New York: Bantam Books, First Vintage Books Edition, 2005), 331.

24   Greene, *The Fabric Of The Cosmos*, 257.

any other form of information existed prior to the big bang, it would have "melted" in the big bang, and would not survive the transition across the big bang. (If we accept Hawking's theory that time does not exist prior to the big bang, then in this case we would simply say that no information could exist during the period of the big bang's high temperature.)

Hawking agrees with this statement, because he says:

> It seems the laws of the evolution of the universe may break down at the big bang. If they do, it would make no sense to create a model that encompasses time before the big bang, *because what existed then would have no observable consequences for the present* [italics mine], and so we might as well stick with the idea that the big bang was the creation of the world.[25]

Hawking is saying that anything that existed *before* the big bang would have no observable consequences for anything that occurred after the big bang. This means that we have no way of justifying the statement that virtual particles helped create the baby universes because we have no way of knowing if there were any virtual particles before the big bang. This is the same reservation about "something coming from nothing" that Nobel Prize winning physicist Charles H. Townes had when he said:

> It is true that physicists hope to look behind the "big bang," and possibly to explain the origin of our universe as, for example, a type of fluctuation. But then, of what is it a fluctuation and how did this in turn begin to exist?[26]

---

25 Hawking, *The Grand Design*, 51.
26 Timothy Ferris, *The Whole Shebang* (New York: Simon & Schuster, 1997), 245.

The big bang theory says that all the mass and energy in our universe today was in a tiny space much smaller than an atom. Greene wrote an interesting discussion on how this might have occurred. Greene says:

> If you want a universe like the one we see today, you have to start with raw material whose mass and energy we see today. The big bang theory takes such raw material as an unexplained given ... How do we ignite the bang? ... With a powerful outward burst of spatial expansion being the trade mark, the inflationary theory puts a bang in the big bang, and a big one at that; according to inflation, an anti-gravity blast is what set the outward expansion of space in motion ... The only independent energy budget required by inflationary cosmology is what's needed to create an initial inflationary seed, a small spherical nugget of space filled with a high-valued inflaton field that gets the inflationary expansion rolling in the first place. When you put in numbers, the equations show that the nugget need be only about 10^{-26} centimeters across and filled with an inflaton field whose energy, when converted to mass, would weigh less than ten grams ... That's a lump you could put in your wallet.[27]

An inflaton field is what produces inflation, although we do not yet know what an inflaton field is.

---

27   Brian Greene, *The Hidden Reality* (New York: First Vintage Books Edition, 2011), 316–318.

Very Small Is Not Nothing

Greene is saying that in his picture of how the universe began from nothing, you do need to start with *something*. But that something is so small and innocuous that you could put it in your wallet! In other words, Greene is essentially saying that we may ignore the question of how this initial lump got there because it is so small. But the problem with this assumption is that the size and weight of this lump, regardless of how small, are relative to the size of its surroundings at the initial time of the big bang. The small lump is enormous compared to the dimensions we are dealing with at the time of the big bang. Greene says that the size of the small lump would be 10^{-26} centimeters across. Scientists estimate that initially the big bang had a dimension smaller than 10^{-33} centimeters across, less than a Planck unit of distance. This means that the size of the lump that can fit in your wallet is anything but negligible compared to the initial size of the big bang. In fact, the size of the small lump is ten million times larger than the initial size of the big bang (10^{-26} divided by 10^{-33} equals 10^7, or ten million).

To *assume* the preexistence of the small lump (that is ten million times larger than the original size of the big bang) leaves us with a new question. How did this little lump that could fit in your wallet get there in the first place? This is at least as problematic a question as how the big bang got there.

There is also this question: Where did the inflaton field come from? Greene admits that this is a challenge. Greene says:

> This approach, nevertheless, presents daunting challenges.
> For one thing, the inflaton remains a purely hypothetical
> field. Cosmologists freely incorporate the inflaton field
> into their equations, but unlike electron and quark fields,

there is as yet no evidence that the inflaton field exists. For another, even if the inflaton proves real, and even if we one day develop the means to manipulate it much as we do the electromagnetic field, still the density of the requisite inflaton seed would be enormous: about 10^{67} times that of an atomic nucleus. Although the seed would weigh less than a handful of popcorn, the compressive force we would need to apply is trillions and trillions of times beyond what we can now muster.[28]

In Greene's picture of the origin of the universe there remains the problem of where the inflaton field came from, in addition to where the wallet lump came from. It may turn out that in the years to come, the theory of inflation will have answers to these questions, and scientists may accept eternal inflation as a well-established scientific theory, but we are not yet at that point.

The Existence of Gravity

In order to explain his theory of the origin of the universe, Hawking assumes the existence of gravity. Hawking says:

> If the total energy of the universe must always remain zero, and it costs energy to create a body, how can a whole universe be created from nothing? That is why there must be a law like gravity. Because gravity is attractive, gravitational energy is negative: one has to do work to separate a gravitationally bound system. This negative energy can balance the positive energy needed to create

---

28   Greene, *The Hidden Reality*, 318.

matter ... Because there is a law like gravity, the universe
can and will create itself from nothing.[29]

Hawking is assuming that the law of gravity *already* existed when
the quantum fluctuations created the baby universes. He says that
the reason the universe can create itself is the existence of the law of
gravity ("Because there is a law like gravity, the universe can and will
create itself from nothing"). But Hawking gives no explanation of how
the law of gravity came to exist *before* anything else existed. This is a
major obstacle to Hawking's explanation of how the universe created
itself. There is no scientific reason why one particular law such as the
law of gravity should come into existence before anything else existed.
It is just as big a problem to explain how gravity came into existence
as to explain how the universe came into existence. To say that gravity
already existed when the universe self-created would make as much
sense as saying: Because there is a law like gravity, the law of gravity
can and will create itself.

There may have been laws of physics before the big bang, but we
have no way of knowing if there were, or what they were, *because no
information could get across the big bang temperature barrier.*

Singularities: Yes or No?

Earlier in his career (1970), Hawking believed that the origin of
the universe came about from a big bang *singularity*. Mathematicians
call a situation where something is equal to infinity a singularity. Many
scientists do not like working with infinite amounts. A singularity at the
big bang means all the matter existent at that time was compressed into
a region of zero volume, so that the density of matter, the curvature of
space-time, and the temperature, became infinite. It is generally accepted

---

29   Hawking, *The Grand Design*, 180.

by scientists that if infinity appears in some equation, something is wrong with that equation. Hawking worked on this problem for years. By the time Hawking wrote *The Grand Design* in 2010, he had changed his mind, as he says:

> The final result was a joint paper by Penrose and myself in 1970, which at last proved that there must have been a big bang singularity provided only that general relativity is correct and the universe contains as much matter as we observe. There was a lot of opposition to our work, partly from the Russians because of their Marxist belief in scientific determinism, and partly from people who felt that the whole idea of singularities was repugnant and spoiled the beauty of Einstein's theory. However, one cannot really argue with a mathematical theorem. So in the end our work became generally accepted and nowadays nearly everyone assumes that the universe started with a big bang singularity. It is perhaps ironic that, having changed my mind, I am now trying to convince other physicists that there was in fact no singularity at the beginning of the universe.[30]

Although Hawking no longer believes that there was a singularity at the big bang, he does believe there was a tremendous amount of energy present in order to kick-start the expansion of the universe. How did that much energy get there at the beginning of the big bang? Hawking asks: "Why was the early universe so hot?" Hawking acknowledges that this question is still unanswered.[31]

And elsewhere Hawking says:

30   Hawking, *A Brief History of Time*, 50.
31   Hawking, *A Brief History of Time,* 121.

If we extrapolate to the distant past, all the matter and energy in the universe would have been concentrated in a very tiny region of unimaginable density and temperature, and if we go back far enough, there would be a time when it all began—the event we now call the big bang.[32]

The Inflation Theory

What suddenly caused the energy to stop rising and explode into the universe? Most cosmologists today would answer that the *inflation theory* is involved in the early behavior of our universe. Inflation theory was originally proposed in the early 1980s by Alan Guth, a physicist at MIT. A greatly simplified explanation of inflation theory is as follows: First there was the big bang, and about 10^{-35} seconds later, inflation began, brought on by an *inflaton field*. In a small fraction of a second, the universe expanded from about the size of a proton to the size of a grapefruit,[33] a multiplication in size of about 10^{25}. At a time less than one second later, inflation stopped and normal expansion was restored.

Hawking says about inflation:

> It was as if a coin 1 centimeter in diameter suddenly blew up to ten million times the width of the Milky Way. That may seem to violate relativity, which dictates that nothing can move faster than light, but that speed limit does not apply to the expansion of space itself.[34]

Hawking then talks about the big bang and inflation:

---

32 Hawking, *The Grand Design*, 124.
33 Davies, *The Goldilocks Enigma*, 56.
34 Hawking, *The Grand Design*, 129.

Inflation explains the bang in the big bang, at least in the sense that the expansion it represents was far more extreme than the expansion predicted by the traditional big bang theory … The problem is, for our theoretical models of inflation to work, the initial state of the universe had to be set up in a very special and highly improbable way. Thus traditional inflation theory resolves one set of issues but creates another—the need for a very special initial state. That time-zero issue is eliminated in the theory of the creation of the universe we are about to describe.[35]

The No-Boundary Condition

Hawking believes that a *no-boundary condition* universe would resolve the problem of needing very special initial conditions. It also would solve the problem of how the beginning of the universe occurred. Hawking said:

So long as the universe had a beginning, we could suppose it had a creator. But if the universe is really completely self-contained, having no boundary or edge, it would have neither beginning nor end: it would simply be. What place then for a creator?[36]

Hawking says that the quantum theory of gravity has opened up the possibility that there is no need to talk about what happened at the boundary of the origin because there is no boundary. This means there would be no singularity at the big bang. The universe has no origin or beginning, it simply *is*. Of course this depends on one's definition of the word *is*. Hawking says:

35 Hawking, *The Grand Design*, 130–131.
36 Hawking, *A Brief History of Time*, 140, 141.

The quantum theory of gravity has opened up a new possibility, in which there would be no boundary to space-time and so there would be no need to specify the behavior at the boundary. There would be no singularities at which the laws of science broke down and no edge of space-time at which one would have to appeal to God or some new law to set the boundary conditions for space-time. One could say: "the boundary condition of the universe is that it has no boundary." The universe would be completely self-contained and not affected by anything outside itself. It would neither be created nor destroyed. It would just BE.[37]

When one combines the general theory of relativity with quantum theory, the question of what happened before the beginning of the universe is rendered meaningless ... The realization that time behaves like space presents a new alternative. It removes the age-old objection to the universe of having a beginning, but also means that the beginning of the universe was governed by the laws of science and doesn't need to be set in motion by some god.[38]

The no-boundary condition permits Hawking to eliminate the question "How did everything begin?" Hawking says the question "How did everything begin?" is an irrational question because the no-boundary condition says that there is no beginning of time.

There are two serious problems with Hawking's no-boundary theory. Firstly, Hawking derives his no-boundary condition theory from the idea that gravity warps time and space. This idea derives from

---

37   Hawking, *A Brief History of Time*, 136.
38   Hawking, *The Grand Design*, 135.

the theory of general relativity, and the theory of gravity. But the theory of general relativity is not valid when we are trying to determine the behavior of particles at the Planck level (10^{-33} cm), which is what we are talking about at the big bang event, because we know that the whole universe was compressed down to Planck size. (This is the same logic that Hawking used for concluding that there was no singularity at the big bang; because Einstein's theory of general relativity does not work for quantum distances.) This means that we cannot use the general theory of relativity to obtain information about the big bang at the time of the big bang.

Hawking agrees that we cannot use the general theory of relativity to obtain information about the big bang at the time of the big bang, for he says:

> Although one can think of the big bang picture as a valid description of early times, it is wrong to take the big bang literally, that is, to think of Einstein's theory as providing a true picture of the origin of the universe. That is because general relativity predicts there to be a point in time at which the temperature, density, and curvature of the universe are all infinite, a situation mathematicians call a singularity. To a physicist this means that Einstein's theory breaks down at that point and therefore cannot be used to predict how the universe began, only how it evolved afterward. So although we can employ the equations of general relativity and our observations of the heavens to learn about the universe at a very young age, it is not correct to carry the big bang picture all the way back to the beginning.[39]

---

39   Hawking, *The Grand Design*, 128, 129.

At the time of the big bang we are dealing with gravity on the Planck scale (10^{-33} cm), hence *we need a quantum theory of gravity.* Yet scientists are unanimous in agreeing that *we do not yet have a quantum theory of gravity.* The recent discovery of the *Higgs boson* is breathtakingly exciting. It helps open the door a little further to many scientific mysteries that still remain. The recent discovery of the Higgs boson may help scientists to find a quantum theory of gravity, and to solve other present day mysteries in science.[40]

A No-Boundary Condition Requires a Quantum Theory of Gravity

Hawking takes what we *do* know about quantum mechanics and applies it to the time of the big bang, thus arriving at the conclusion that he calls the no-boundary condition. But the problem with doing this is that some of the yet unknown portions of a quantum theory of gravity apply to the time of the big bang, and this may yield results that are quite different from Hawking's no-boundary condition. It is therefore an enormous assumption to derive a theory like the no-

---

40   Sean Carroll, a theoretical physicist at Caltech, wrote an interesting article for the CNN website titled "How the Higgs can lead us to the dark universe." On July 24, 2012, Carroll said, "The incredible discovery of the Higgs boson will open up new ways of probing the part of the universe that is invisible to our everyday senses: beyond ordinary matter, into the extraordinary world of dark matter ... Part of the excitement stems from the fact that the Higgs boson is the final piece in an extremely elaborate puzzle: the Standard Model of particle physics ... With this final piece in place, we can justifiably say that we understand the behavior of ordinary matter—the atoms and molecules that make up ourselves and our everyday world ... As successful as the Standard Model has been, we know it's not the final answer to how the universe works. Strong evidence comes from the existence of dark matter: mysterious, invisible stuff that adds up to five times as much mass as the ordinary atoms and particles in the universe." http://www. cnn.com/2012/07/20/opinion/higgs-dark-matter-carroll/index.html.

boundary condition by using an *incomplete* quantum theory! Our present quantum theory is incomplete because it does not include a quantum theory of gravity. So it is not possible to say that the no-boundary condition derives from the laws of physics, because the laws of physics are incomplete, up to this time, with regard to the quantum theory of gravity. One of the best-known science writers, Timothy Ferris, refers to this in his chapter, "The Origin of the Universe", where he looks at the leading theories of the universe's origin and says:

> It is perhaps unnecessary to caution that the Hartle-Hawking wave function does not explain the origin of the universe … None of the other theories discussed in this chapter does either. All are actually rather limited. They omit quantum gravity, for which there is as yet no realized theory.[41]

We have seen that the no-boundary condition cannot be called a scientific theory, because this would require a quantum theory of gravity, which we do not yet have. So the no-boundary condition can at best be called a conjecture. Hawking admits that his no-boundary theory is simply a *proposal*, and it is not derived from scientific equations. Hawking says:

> I'd like to emphasize that this idea that time and space should be finite without boundary is just a *proposal*: it cannot be deduced from some other principle.[42]

In order to look more closely at what happened during the big bang, we need a theory of quantum gravity. Brian Greene has stated the situation succinctly:

---

41   Ferris, *The Whole Shebang*, 253.
42   Hawking, *A Brief History if Time*, 136.

We have made great strides in piecing together a consistent and predictive story of cosmic evolution. But the picture remains incomplete because of the fuzzy patch near the inception of the universe ... If we ever hope to understand the origin of the universe—one of the deepest questions in all of science—the conflict between general relativity and quantum mechanics must be resolved. We must settle the difference between the laws of the large and the laws of the small and merge them into a single harmonious theory.[43]

Here is an additional problem with Hawking's no-boundary condition. According to Hawking's no-boundary condition, it is meaningless to ask what happened before the beginning of the universe, because there is no *time*, as we know it, before the beginning of the universe. And yet Hawking says the big bang is the beginning of the universe. But how can the big bang occur as a result of what happens *before* the big bang? Remember, Hawking said, "Quantum fluctuations lead to the creation of tiny universes out of nothing." How can there be a big bang if it comes from quantum fluctuations that occur when time does not exist? These quantum fluctuations must have occurred *before* the big bang if it was the quantum fluctuations that produced the big bang. But if there is no time before the big bang, how can there have been quantum fluctuations that produced the big bang? So the idea of a no-boundary condition does not seem consistent with the idea that quantum fluctuations produced the big bang. Furthermore, Hawking agrees that having a quantum theory of gravity is essential if we are to discuss the big bang. Hawking says:

If we want to understand the early universe, when all the matter and energy in the universe were squeezed into a

---

43   Greene, *The Fabric of the Cosmos*, 337–338.

small volume, we must have a quantum version of the theory of general relativity. We also need such theories because if we are seeking a fundamental understanding of nature, it would not be consistent if some of the laws were quantum while others were classical.[44]

While on the one hand, Hawking admits that a quantum theory of gravity is necessary if we are to discuss the early universe, on the other hand, he proceeds to discuss the big bang and the no-boundary condition without this necessary quantum theory of gravity.

If God Created The Universe, Who Created God?

Hawking discusses the possibility of the existence of God. Hawking says:

It is reasonable to ask who or what created the universe, but if the answer is God, then the question has merely been deflected to that of who created God.[45]

Hawking is saying that part of the reason he cannot believe in God is because we cannot answer the question "Who made God?" Hawking then turns around and uses the Augustinian theory about who made God to answer the question of how the no-boundary condition came to be. (St. Augustine answered the question of who made God by saying no one made God because God has always existed.) The no-boundary condition of the universe means the universe had no beginning and no end. It simply *is!* Why cannot this same logic apply to God? He had no beginning and no end. He simply *is*. Hawking is willing to believe that something can always have existed when it comes to his no-boundary condition, but not when it comes to God!

---

44   Hawking, *The Grand Design*, 103.
45   Hawking, *The Grand Design*, 172.

So Far, All the Theories of Creation from Nothing Involve Creation from Something

All the various theories about how something can come from nothing assume the preexistence of at least some of the laws of physics, and some form of matter, such as virtual particles. Hawking said: "Because there is a law like gravity, the universe can and will create itself from nothing." But none of these theories explains how the law of gravity or the laws of physics arose! Hawking never attempts to explain how gravity came to preexist the big bang. In fact, according to Hawking's no-boundary condition, gravity (or the law of gravity) could not have existed before the big bang, because there was nothing before the big bang, and Hawking says it makes no sense to ask any question about before the big bang. We are then left with the supposition that in Hawking's view, gravity (or the law of gravity) must have come into existence *during* the bang, which is hard to imagine, given the temperature during the big bang.

Lawrence Krauss has written a book titled *A Universe from Nothing*. Krauss believes that something has come from nothing. Krauss says:

It certainly seems sensible to imagine that a priori, matter cannot spontaneously arise from empty space, so that *something*, in this sense, cannot arise from *nothing*. But when we allow for the dynamics of gravity and quantum mechanics, we find that this commonsense notion is no longer true ... While inflation demonstrates how empty space endowed with energy can effectively create everything we see ... it would be disingenuous to suggest that empty space endowed with energy, which drives inflation, is really *nothing. In this picture one must assume that space exists and can store energy* [italics mine] and one uses the laws of physics like general relativity to calculate the consequences.[46]

Krauss admits "one must assume that space exists and can store energy, and one uses the laws of physics."

Then Krauss says:

> The question of what determined the laws of nature that allowed our universe to form and evolve now becomes less significant. If the laws of nature are themselves stochastic and random, then there is no prescribed "cause" for our universe. Under the general principle that anything that is not forbidden is allowed, then we would be guaranteed, in such a picture, that some universe would arise with the laws that we have discovered. No mechanism and no entity is required to fix the laws of nature to be what they are. They could be almost anything.[47]

Krauss is saying here that we need not be concerned with how the laws of physics came to be, since with an infinite number of universes there will be one that has the same laws as ours. But there is a problem with Krauss's view. According to those who believe in an infinite number of universes, at the time of the origin of our universe there did not yet exist an infinite number of universes. The infinite number of universes came into being as the end result of a process of *replication* of universes, which we have referred to elsewhere in this book. So this still leaves us with the question of where the laws of physics came from that helped in the formation of our own universe.

In further addressing himself to this question, Krauss says:

---

46 Lawrence Krauss, *A Universe From Nothing* (New York: Free Press, 2012), 151–152.

47 Lawrence Krauss, *A Universe From Nothing* (New York: Free Press, 2012), 176.

We generally assume that certain properties, like quantum mechanics, permeate all possibilities. I have no idea if this notion can be usefully dispensed with, or at least I don't know of any productive work in this regard.[48]

It is not the objective of this chapter to show that we can never discover such a theory, rather to show that we have *not yet* reached the day when we can say the laws of physics show us that something can come out of nothing.

Should a quantum theory of gravity be discovered, and Hawking's arguments that physics shows how something can come from nothing are found to be scientifically valid, there still remains a far larger problem with Hawking's statement that "the laws of physics show that God is unnecessary." That larger problem is this: How do we account for Hawking's Apparent Miracle, the exceptional fine-tuning of our home planet? The next two chapters focus on this question.

---

48   Krauss, *A Universe From Nothing*, 177.

CHAPTER 4

The Mother of All Physics Problems

ACCORDING TO HAWKING, THE biggest Apparent Miracle is the value necessary for the cosmological constant in Einstein's equations to result in a universe that is expanding at the rate of our own universe, and that makes it possible for human life to exist on earth. Hawking says:

> The most impressive fine-tuning coincidence involves the so-called cosmological constant in Einstein's equations of general relativity.[49]

If the cosmological constant were a tiny bit larger or smaller, we would not have a universe that has existed for 13.7 billion years, thereby giving the stars enough time to produce the heavy elements, and giving enough time for animals and humans to evolve on earth. The cosmological constant, also referred to as *vacuum-energy* or *dark energy*, is the term in Einstein's equations that influences whether the universe is contracting or expanding, and at what rate this expansion or contraction is taking place.

The cosmological constant is the famous term that Einstein referred to as the "greatest blunder" in his career. Einstein referred to

---

49   Hawking, *The Grand Design*, 161.

it as his blunder because in 1917 scientists believed in a static universe, that is, that our universe has existed forever and is neither expanding nor contracting. So Einstein created a cosmological constant that he inserted into his equations on relativity. He assigned to this cosmological constant a value that would have an expansive influence on matter to precisely counteract the attractive force between the elements of space, thus allowing the universe to be neither contracting nor expanding. It was not until twelve years later (in 1929) that Hubble showed that the universe is not static, but it is expanding. At that point Einstein gave up on pursuing the antigravity effect of the cosmological constant.

A troubling question remained. If the only gravitational force exerted on matter is by Newton's law of gravity, there is no way that the universe could continue to exist because the universe would implode upon itself. When it was discovered that our universe is actually increasing in size, then scientists began to explore this unanswered question. The answer is that there must be some kind of force at work that we have not yet discovered, some antigravity force or cosmic repulsive force. But this is a totally unexpected result. It means that there is some invisible force in the universe that does not merely prevent the universe from imploding upon itself, but even causes the universe to *expand*. It is a force that, to this date, scientists have been unable to identify, and whose source remains a total mystery. This has come to be called *dark energy*. So today scientists are convinced that Einstein's equations do need a cosmological constant, which is equivalent to the dark energy that is causing the universe to expand. And it is really dark—so dark we cannot see it, we do not know what it is, and we do not know where it comes from or how it got wherever it is!

Hawking has some fascinating comments on present studies of the cosmological constant. Hawking says:

Physicists have created arguments explaining how it (the cosmological constant) might arise due to quantum mechanical effects, but the value they calculate is about 120 orders of magnitude (a 1 followed by 120 zeros) stronger than the actual value, obtained through supernova observations. That means that either the reasoning employed in the calculation was wrong or else some other effect exists that miraculously cancels all but an unimaginably tiny fraction of the number calculated. The one thing that is certain is that if the value of the cosmological constant were much larger than it is, our universe would have blown itself apart before galaxies could form and life as we know it would be impossible.[50]

Estimates of the dark energy in our universe can be made through finding the vacuum energy stored in electrons, gravitons and the other known particles. But the answers come out so large that the gravitational repulsion would destroy galaxies, atoms, and our universe. Why has this not happened? Susskind says:

We have sought after such an explanation for almost half a century with no luck … Finding the reason has been regarded as the biggest, most important, and most difficult problem of modern physics. No other phenomenon has puzzled physicists for as long as this one. Every attempt, be it in quantum field theory or in String theory, has failed. It is truly the mother of all physics problems.[51]

What makes this "the mother of all physics problems?" It is because of the odds against the earth having the right cosmological constant by

---

50   Hawking, *The Grand Design*, 162.
51   Susskind, *The Cosmic Landscape*, 78.

chance. Scientists today are generally agreed that if the cosmological constant were just an order of magnitude bigger or smaller than 10^{-120}, then no stars or planets that endured for billions of years would have been formed. This means it would not be possible for human life to exist, since most of the chemicals that comprise the earth and human beings come from inside the furnaces of stars.[52]

The number 10^{-120} is important because it gives us an idea of the limits of the probability that the value of the cosmological constant occurred by chance. In other words, the probability of the odds that the earth has a cosmological constant (as a result of sheer chance) that allows for life must be in the vicinity of 10^{120} to one against it. Susskind says:

> For a bunch of numbers none of them particularly small, to cancel one another to such precision would be a numerical coincidence so incredibly absurd that there must be some other answer.[53]

The importance of this statement is that Susskind is saying that odds of 10^{120} to 1 are so unlikely, that it is unreasonable to say that something with such high odds against it happened by chance. We must look for some other solution!

That our cosmological constant has a value that permits the existence of stars and planets and galaxies for billions of years does not mean that the odds against *life* existing in our universe are 10^{120} to one. Those odds are much larger than 10^{120} to one because many parameters in addition to the cosmological constant enter the picture when it comes to life existing on our planet. (See the quotation above by Davies about the Designer Machine and the thirty knobs.) *The*

---

52   Susskind, *The Cosmic Landscape*, 78–85, has a good discussion of this.
53   Susskind, *The Cosmic Landscape*, 78.

cosmological constant is only one of the thirty knobs. So the actual odds against our universe having all the parameters necessary by chance will be much larger than 10^{120} to one, when we take into account the odds against each of the other twenty-nine knobs being set by chance. Our universe has existed for 13.7 billion years. Earth has existed for about 4.5 billion years. The cosmological constant has an extremely narrow range for permitting our universe and earth to exist for the times they have existed, but we will have to figure out the odds of the other twenty-nine knobs before we know the odds against the existence of an earth that can support life. Doing so is probably not possible at this stage of our knowledge, since there are too many unknowns in the question of how the other twenty-nine knobs interact with each other, if at all, when any one knob is altered by a small amount. (But there is another approach to this question in chapter 6.)

The mother of all physics problems is as follows: How do we account for the fact that our universe has a cosmological constant that is so fine-tuned that it would require transcending odds of 10^{120} to one to hit the precise value for our cosmological constant that allows human life to exist?

CHAPTER 5

Would 10^{500} Universes Be Enough?

HAWKING MADE A PROPOSAL that would overcome the odds of 10^{-120} to one. In order to solve the Apparent Miracle, Hawking calls on M-theory. Hawking accepts the M-theory's finding that there can be up to 10^{500} different universes, each with different laws. Hawking says:

> The fine tunings in the laws of nature can be explained by the existence of multiple universes.[54]

If there are 10^{500} universes, then Hawking thinks it almost certain that at least one of them would have the cosmological constant of our universe, and thus make it possible for life to exist on our planet. Susskind agrees:

> With that many possibilities to choose from, it is overwhelmingly likely that the energy of many vacuums will cancel to the accuracy required by Weinberg's anthropic argument, namely 119 decimal places (10^{120}).[55]

Hawking and Susskind are saying the number 10^{120} is so much smaller than 10^{500} that we will find it overwhelmingly likely that

---

54   Hawking, *The Grand Design*, 165.
55   Susskind, *The Cosmic Landscape*, 109.

many of the universes in the 10^{500} will match the vacuum number (cosmological constant) of the one in 10^{120}. This means that many of the universes out of the 10^{500} will have a planet with a cosmological constant that would allow the existence of a planet like our earth. This is a mathematically sound statement, if we are taking the two principal numbers to be 10^{120} and 10^{500}.

But Hawking has made a mistake here. He bypasses the fact that the odds against Hawking's Apparent Miracle are actually much larger than 10^{120} to one. The number 10^{120} represents only *one* of the thirty knobs that go into making our planet livable for human beings. All thirty knobs must be tweaked in order for earth to have the characteristics that would allow for life. So we need a calculation that gives us the odds of producing a universe that has a planet with all thirty knobs of the Designer Machine suggested by Davies, tweaked in a manner that allows life to exist.

The Roger Penrose Number

Roger Penrose, one of the world's most honored scientists, has calculated the mathematical odds of our universe meeting the circumstances of Hawking's Apparent Miracle by pure chance.

Penrose has received a number of awards and prizes, including the Albert Einstein Award, and the 1988 Wolf Prize, which he shared with Stephen Hawking for their understanding of the universe. He was knighted in 1994 for his service to science.

Penrose approaches the probability of our universe existing by pure chance from a much different point of view—the point of view of the total entropy in the universe. He arrives at a staggering figure for the probability of a universe like ours being formed by chance: the odds are 10 to the $10^{(123\ zeros)}$ to one against it![56]

In order to give you a feeling for the comparison of these two important numbers, I will write them out.

The number of universes suggested by M-theory, according to Hawking is as follows:

(A) 10^{500}

The probability of a universe such as ours existing by random chance, according to Penrose, is one in:

(B) $10^{100000000000000000000000000000\ 0000000000000000000000000000000000000000000}$ 00000000000000000000000000000000000000000000000000000

There are 123 zeros in the exponent above. You don't have to count them—unless you really want to!

When you compare the two numbers above it is clear that 10^{500} different universes is not nearly enough to insure that at least one of these 10^{500} universes will have the necessary parameters for life on earth to exist, because the odds against having parameters for life to exist by chance is the number (B). The number I have designated as (A) would need to be millions of *orders of magnitude* larger than the number designated as (B). The number (A) is breathtakingly far from being larger than (B). In fact, the number (B) is stunningly larger than (A). The number (B) is so much larger than (A) that it almost makes (A) look like zero!

If M-theory is correct, and there are about 10^{500} universes, that

---

56   Roger Penrose, *The Emperor's New Mind*, (Oxford: Oxford University Press, 1989), 440–445. This shows where you may follow the mathematics of how Penrose obtained the number 10 to the $10^{(123\ zeros)}$. His approach is based on the total entropy of the universe. It is ironic that in doing this, Penrose uses a formula proposed by Stephen Hawking dealing with the entropy of a black hole. (Stephen Hawking, "Particle creation by black holes", *Commun. Math. Phys.*, 43, 199, 200.)

is nowhere near enough to ensure that the mathematical odds will be met for allowing life to exist on one of these 10^{500} universes. These two numbers show it is highly unlikely that M-theory (with its 10^{500} universes) could explain how our universe originated. These two numbers show it is mathematically highly improbable for our universe to fulfill Hawking's Apparent Miracle by a chance occurrence.

The question was posed at the beginning of this section: *Is* 10^{500} universes enough? The answer is no. It is nowhere near enough!

The number calculated by Penrose is enough in itself to cast doubt upon the statement of Hawking that the laws of physics show that it is unnecessary to believe in a supernatural agent of creation.

I am not saying that this *proves* that Hawking is wrong, but I *am* saying that mathematical probabilities are strongly counter to Hawking's statement.

Also, please note that this does *not* say that the Penrose number shows that there *is* a supernatural agent of creation. *It simply says it is not possible to say that the laws of physics in Hawking's multiverse show that God is unnecessary.*

CHAPTER 6

Would an Infinite Number of Universes Be Enough?

HAWKING DOES NOT BELIEVE there are an infinite number of universes. Hawking says:

> M-theory predicts that a great many universes were created out of nothing ... They are a prediction of science. Each universe has many possible histories and many possible states at later times ... Most of these states will be quite unlike the universe we observe and quite unsuitable for the existence of any form of life. Only a very few would allow creatures like us to exist. Thus our presence selects out from this vast array only these universes that are compatible with our existence.[57]

Hawking's words "only a very few" make it clear that he does not believe there are an infinite number of universes. If Hawking believed there are an infinite number of universes he would not have used the words "only a very few" in talking about how many universes exist that are compatible with our existence, because if there are an infinite

---

57   Hawking, *The Grand Design*, 9.

number of universes, then there would be an *infinite* number of planets just like earth, not only a very few.

Since there are scientists who do believe there are an *infinite* number of universes, it is important to discuss that theory in this chapter. Would an infinite number of universes be a large enough number to insure that at least one of these universes would include a planet that meets the conditions required for life to exist, and therefore explain Hawking's Apparent Miracle?

Yes.

An infinite number of universes would be a sufficient number of universes to support the claim that the odds favor at least one of these universes having a planet with parameters that would support life as we know it on our planet. If there are an infinite number of universes, then Hawking's Apparent Miracle (and the Goldilocks Enigma) disappear as problems. In fact, if there are an infinite number of universes, *there will be an infinite number of planets that are precisely like earth*! This is so because in the strange mathematical world of infinity if you take infinity, and divide it by any number less than infinity, the result is still infinity. Infinity is a very large number!

Brian Greene believes that there are an infinite number of universes. According to Greene, some of these universes will have people identical to you and me. (In fact, there will be an infinite number of such people!) Greene says:

> In the far reaches of an infinite cosmos, there's a galaxy that looks just like the milky way, with a solar system that's the spitting image of ours, with a planet that's a dead ringer for earth, with a house that's indistinguishable from yours, inhabited by someone who looks just like you, who is right now reading this very book and imagining

you, in a distant galaxy, just reaching the end of this sentence. And there's not just one such copy. *In an infinite universe there are infinitely many.* [Italics mine] In some, your *doppelganger* is now reading this sentence, along with you. In others, he or she has skipped ahead, or feels in need of a snack and has put the book down. In others still, he or she has, well, a less than felicitous disposition and is someone you'd rather not meet in a dark alley.[58]

The quotation above by Greene is true in a cosmos with an infinite number of universes. Such are the staggering implications of postulating an infinite number of universes. As was said earlier, infinity is a very large number. *Really* large!

Expanding upon Greene's words, if there are an infinite number of universes, then there is a house in one of these universes that's indistinguishable from yours, there is another house exactly like yours except that it has two bedrooms instead of three, there is another one with a Rolls-Royce in the garage instead of a Ford, there is another one with a mango tree in the backyard instead of an apple tree, there is another one with a person just like you, except this person has garters on his socks instead of no socks, and so on *ad infinitum*! You get the idea.

The only way there can be an infinite number of universes in the present is if the first universe or universes began in the infinite past. If the origin of the universe occurred a *finite* time ago, then not enough time would have expired between the origin of the universes and the present time for the "replication process of universes"[59] (under the eternal inflation theory) to produce an infinite number of universes. It

---

58   Greene, *The Fabric Of The Cosmos*, 11.
59   Greene, *The Hidden Reality*, 62, 63; a discussion of how universes continue to be born anew in the *eternal inflation theory*.

can be argued convincingly that an infinite past or infinite future is a point in time that can never be actually reached. These are *imaginary* points in time. For example, the moment we think we have reached a particular point infinitely into the future, we have not really reached it, because we could still go further into the future!

On the other hand, if the universe started in the infinite past, this universe (and its children) will never reach our present day, for how would one describe the transition point or time, when the infinite past moved onto the axis of finite measurable time? Infinity is not a real number in the sense that we could ever write down precisely what it is. It is a mathematical concept that can be helpful to us in many circumstances, but when we talk about something taking place in the infinite past, we have put it outside of our reach. (*Infinitely* outside of our reach!)

According to the *eternal inflation theory*, an infinite number of universes would exist only at the end of the infinite period of time, meaning an infinite time into the future.

Greene does believe in an eternal past, for he says:

> If space is truly infinite in size, then it always has been and always will be.[60]

Not all cosmologists would agree with Greene on this point. Alan Guth, considered the "father of inflation theory," presents a mathematical proof that our universe does not have an infinite past, in an article published by MIT. Guth says:

> Although inflation is generically eternal into the future, it is not eternal into the past: it can be proven under reasonable assumptions that the inflating region must be incomplete in past directions, so some physics other than

60   Greene, *The Hidden Reality*, 30.

inflation is needed to describe the past boundary of the inflating region.[61]

Paul Davies approaches the question of an infinite past from a different perspective, and has written an excellent discussion on why the universe could not have existed forever.[62]

It is highly unlikely that there exists something like a multiverse with an infinite number of universes, since there has never existed an infinite number of *any* material item. There are many infinite series of numbers, but that is quite different from saying that there are an infinite number of any material object. You cannot write on a piece of paper all the individual numbers represented in any infinite series of numbers. So we say again: There has never existed an infinite number of any material item. And it is highly unlikely that there exist an infinite number of universes. For us to say an infinite number of universes exist would require some sort of experimental evidence. To date, there is no such evidence. The idea of an infinite number of universes could presently be labeled only as a speculation, not a scientific theory.

Cosmologist George F. R. Ellis discusses the challenge of multiverses. Ellis is considered one of the world's leading experts on Einstein's general theory of relativity and is coauthor, with Stephen Hawking, of *The Large Scale Structure of Space-Time*. Ellis has also written an interesting article in which he concludes that:

> Parallel universes may or may not exist; the case is unproved.
>
> We are going to have to live with that uncertainty.

---

61   Alan H. Guth, "Eternal inflation and its implications" (Cambridge: MIT Center for Theoretical Physics, 2007), http://arxiv.org/pdf/hep-th/0702178v1.pdf. Guth's discussion, including his mathematical proof that our universe does not have an infinite past, is in this article that may be downloaded from the website.

62   Davies, *The Goldilocks Enigma*, 71.

Nothing is wrong with scientifically based philosophical speculation, which is what multiverse proposals are. But we should name it for what it is.[63]

It seems unlikely we can ever establish a theory of an infinite number of universes. The occurrence of infinity in a theoretical equation that would result in an infinite number of universes is not possible according to mathematics. Hawking believes that as soon as we talk about something infinite in an equation, that equation breaks down. Hawking says:

> General relativity predicts there to be a point in time at which the temperature, density, and curvature of the universe are all infinite, a situation mathematicians call a singularity. To a physicist this means that Einstein's theory breaks down at that point and therefore cannot be used to predict how the universe began, only how it evolved afterward.[64]

Another reason it is unlikely that we can ever establish that there are an infinite number of universes is their distance from us. There can never exist any kind of communication between any other universe and our universe. George Ellis explains some of the problems we face in trying to "see" what is outside of our own universe.

> When astronomers peer into the universe, they see out to a distance of about 42 billion light-years, our cosmic horizon, which represents how far light has been able to

---

63 George F. R. Ellis, "Does the Multiverse Really Exist?" (Scientific American, Aug. 2011), 43.

64 Hawking, *The Grand Design*, 129. Paul Davies agrees, saying: "When a physical theory contains an infinite quantity, the equations break down and we cannot continue to apply the theory." See Davies, *The Goldilocks Enigma*, 68.

travel since the big bang (as well as how much the space of the universe has expanded in size since then) … no possible astronomical observations can ever see those other universes.[65]

It's Either an Infinite Number of Universes, or God!

There is at least one scientist who states that we need to accept an *endless landscape*, because the only other recourse is to accept *supernatural forces*. Leonard Susskind says:

> What are the alternatives to the populated Landscape paradigm? My own opinion is that once we eliminate supernatural agents, there is none that can explain the surprising and amazing fine-tunings of nature … As I have repeatedly emphasized, there is no known explanation of the special properties of our pocket (universe) other than the populated Landscape—no explanation that does not require supernatural forces.[66]

Highly significant is Susskind's statement that if we set aside supernatural agents, there is no solution that can explain the fine-tuning of our earth, other than that there are an infinite number of universes.

Susskind makes an interesting comment on the process of explaining natural phenomena by anything other than natural phenomena. Susskind says:

---

65   Ellis, Scientific American, Aug. 2011, 38–39.
66   Susskind, *The Cosmic Landscape*, 356, 363. By the terms *populated Landscape*, or *megaverse*, Susskind means an infinite number of universes. "Given a megaverse, endlessly filled with pocket universes …" (363).

But scientists—real scientists—resist the temptation to explain natural phenomena, including creation itself, by divine intervention. Why? Because as scientists we understand that there is a compelling human need to believe—the need to be comforted—that easily clouds people's judgment. It's all too easy to fall into the seductive trap of a comforting fairy tale. So we resist, to the death, all explanations of the world based on anything but the Laws of Physics, mathematics, and probability.[67]

I am in total agreement with Susskind concerning his statement that there are presently only two alternatives to the solution for the problem of the fine-tuning of our universe: either there are an infinite number of universes, or else there is a supernatural force.

Does either of these solutions answer the question of Hawking's Apparent Miracle? Yes. They both do.

Where does that leave us?

---

67   Susskind, *The Cosmic Landscape*, 355.

CHAPTER 7

The Surprising Result of Applying the Laws of Physics and Mathematics to the Origin of Our Universe

THE IMPETUS FOR WRITING this book is Hawking's statement:

> M-theory predicts that a great many universes were created out of nothing. Their creation does not require the intervention of some supernatural being or god. Rather, these multiple universes arise naturally from physical law.[68]

Three unusual events need an explanation in order to confirm Hawking's statement that the laws of physics make it unnecessary to believe in God: the origin of the universe, the origin of life, and the solution to Hawking's Apparent Miracle.

Let me summarize here the reasons given in *Is God Unnecessary?* for doubting Hawking's solution to these three questions.

---

68   Steven Hawking, *The Grand Design*, 8–9.

Nine Reasons Why the Laws of Physics
Do Not Show That God Is Unnecessary.

1. Hawking says we need gravity as a preexistent entity in order to show how our universe came to be out of nothing. How did this preexistent gravity come to be? We have no scientific theory for this.

2. Hawking attempts to postulate a no-boundary condition. He needs a quantum theory of gravity to do this, and we do not have a quantum theory of gravity.

3. Hawking says quantum fluctuations created baby universes. How could quantum fluctuations create the big bang, when nothing existed before the big bang (according to Hawking)?

4. Hawking says that empty space is not really empty; it is filled with virtual particles. But there is no scientific theory to explain how these virtual particles came to be. The measurement of the effects of these virtual particles was made *after* the big bang, not *before* the big bang, when they would be needed to create the big bang. Hawking assumes that virtual particles existed before and after the big bang, but he gives no explanation using the laws of physics as to how the virtual particles survived the astronomical initial temperature of the big bang.

5. Hawking says the Heisenberg uncertainty principle requires that there be virtual particles. But how did the Heisenberg principle come to exist before there were virtual particles? Hawking is saying that three things needed to exist before the big bang: gravity, the Heisenberg Principle, and virtual particles. Even if these three items did exist before the big bang, they would not have survived the high temperature of the big bang (10^{32} kelvin).

6. Hawking poses the question "Why was the early universe so hot?" And he acknowledges that this question is still unanswered.

7. Hawking believes that Charles Darwin showed that the origin of life came about without any supernatural influences, and that this lends support to the theory that the origin of the universe also came about without any supernatural influences. *Charles Darwin himself did not believe this.* Darwin had no scientific theory to account for the origin of life.

8. Hawking's Apparent Miracle is the fine-tuning of our universe to make possible the existence of life on earth. Hawking says that the most amazing fine-tuning factor is the cosmological constant. The odds against the cosmological constant occurring by chance are 10^{120} to one. Hawking says these odds are overcome by the existence of 10^{500} universes. But Hawking errs in saying that the odds against the Apparent Miracle are 10^{120} to one. The cosmological constant comprises only *one* of at least thirty knobs that make up the fine-tuning of the universe according to Davies. So the odds against the fine-tuning of our universe existing by chance are far greater than 10^{120}.

9. Roger Penrose uses a completely different approach to calculate the odds against our specific universe occurring by chance, and the number is more than enormous: 10 to the $10^{(123\ zeros)}$. This supports our statement immediately above that Hawking's number of 10^{500} universes is not nearly large enough to explain the Apparent Miracle.

Here are nine reasons (covered at length in *Is God Unnecessary?*) for why Hawking's understanding of the laws of physics does not explain our existence today. Up to the present date in 2012, the laws of physics do not explain the origin of the universe, the origin of human life, or Hawking's Apparent Miracle (the fine-tuning of our universe).

The main conclusion of Hawking's book, *The Grand Design*, is therefore not valid.

This leaves us with two possible explanations for the problem of the high degree of fine-tuning in our universe:

1. The existence of an infinite number of universes. As we saw earlier, this is not a part of Hawking's proposal, since Hawking believes in the existence of 10^{500} universes.

2. A supernatural creator.

Either one of these two solutions would explain the staggering odds against the fine-tuning of our universe by chance. We may come up with scientific theories in the future that do explain some or all of the nine problems listed above, but we simply do not have them yet. This is in agreement with the statements of Leonard Susskind:

> What are the alternatives to the populated Landscape paradigm? My own opinion is that once we eliminate supernatural agents, there is none that can explain the surprising and amazing fine-tunings of nature … As I have repeatedly emphasized, there is no known explanation of the special properties of our pocket (universe) other than the populated Landscape—no explanation that does not require supernatural forces.[69]

The odds against a chair suddenly jumping one centimeter into the air are about 10^{60} to one (give or take several orders of magnitude). Picture yourself sitting nearby and watching this chair. How long do you think you could sit there before that chair would jump one centimeter into the air? Or would the chair you are sitting on jump up first? Our experience tells us that it would be a long time. A very long time. More than a lifetime. More than a large number of lifetimes. Hawking and Susskind have both stated that the number 10^{120} is so large that if the odds against something happening are 10^{120} to one, then we should look for some possibility other than mere chance. Talking

---

69   Susskind, *The Cosmic Landscape*, 356, 363.

about trying to explain the value of the cosmological constant in terms of a chance occurrence, Hawking says:

> It cannot be so easily explained, and has far deeper physical and philosophical implications.[70]

Where does this leave us? We do not know how to calculate the odds mathematically for or against the existence of an infinite number of universes. We also do not know how to calculate the odds for or against the existence of a supernatural creator. So we are left with the situation where our theories of science and the possibility of the existence of a supernatural creator cannot be favored one over the other by the laws of physics and mathematics.[71]

I personally believe that there is a God. I believe this is compatible with the scientific approach. Some scientists seem concerned that belief in God is a "copout" and that it diminishes our motivation for scientific exploration. Exactly the opposite is true. *Belief in God exalts the scientific quest.* Think of it: a God of infinite IQ has designed an amazing universe, which operates according to extremely complex laws that appear to be applicable to all the stars, planets, and any other matter in our universe. And God has given us human beings the intelligence to discover these laws and examine them. And God encourages us to do so.

---

70   Hawking, *The Great Design*, 162.
71   Davies, "The Goldilocks Enigma", 217 says: "You can't use science to disprove the existence of a supernatural God, and you can't use religion to disprove the existence of self-supporting physical laws." Timothy Ferris says: "we may conclude that atheism is no more soundly footed in cosmological science than is theism … the origin of the universe and of the constants of nature is a mystery, and may forever remain so." (Ferris, *The Whole Shebang*, 310.)

Conclusion

Hawking says that the laws of physics show us that God is unnecessary. In *Is God Unnecessary?* I have listed nine reasons why Hawking's statement is not supportable by the laws of physics as he interprets them.

Where does this leave us as far as science and the existence of God?

It leaves us with a choice between two options:

1. There are an infinite number of universes.

2. There is a God who created the universe.

The laws of physics and mathematics make these two the only choices available. It is the laws of physics and mathematics that show us that the only logical explanation for the fine-tuning of our universe is to say there are an infinite number of universes (in order to overcome the Penrose number), or to say that there is a God who created this fine-tuned universe. The fine-tuning could not have occurred simply by chance.

Which do you choose to believe in: an infinite number of universes, or a God who created the universe and urges us to "meditate upon his works" (an encouragement to pursue scientific exploration)?

Great is the LORD, and greatly to be praised,
and his greatness is unsearchable.
On the glorious splendor of your majesty,
and on your wondrous works, I will meditate.

—Psalm 145:3, 5

Appendix
"The God Experiment."

This appendix describes how I went from agnosticism to a belief in God. Through my undergraduate years at MIT I was skeptical about the existence of God. My parents were not religious or attenders of any church, and I grew up with little interest in God.

While in graduate school I found myself engaged in a discussion with another graduate student in electrical engineering. This discussion was held upon his initiative, not mine. I will summarize the conversation, limiting myself to the salient points.

Jim showed me a passage in the New Testament where Jesus said "Behold, I stand at the door and knock. If anyone hears my voice and opens the door, I will come in to him and eat with him, and he with me" (Revelation 3:20). I looked upon these words of Jesus as an invitation to an experiment.

Up to this point in my life I had spent four years in college studying the scientific method. A part of this scientific method is that in the laboratory we take an equation like F=MA where we can measure the parameters F, M, and A, of a body in motion, insert these parameters into the equation, and thereby verify the validity or falsity of the equation for this particular experiment. There are many theories in science which are so complex that we have not yet figured out how to measure the parameters and insert them into an equation. There are other instances where the theories are so involved that we do not yet have the equation that describes what is taking place, such as the quantum theory of gravity.

In conclusion, one of the aspects of the scientific approach is to take the data into the laboratory, set up the equipment, and try the experiment. It seemed to me that the scientific approach to these words of Jesus was to take the words into the laboratory and "try the experiment."

I did that.

I went into the laboratory and tried the experiment of inviting Jesus Christ into my life. The experiment was successful . Some months later, I remember walking down the hallway to class and thinking to myself "something is different in my life. It feels like some new presence has come into my life that was not there earlier." The presence of God became more and more real to me, and I felt led to attend Seminary in order to learn about the Bible and its historical background, since I had never read any of the Bible up to this point in my life.

What I described above is certainly a subjective experience, but I wanted to explain why I decided to go from science to seminary. Without the years of training in the scientific method at MIT, I might not have been willing to try "the God experiment."

Having said this, I want to emphasize again that the reasoning I have employed in examining Hawking's statement about God is based purely upon the laws of physics, and does not employ any philosophical or theological arguments.

2010760R00040

Made in the USA
San Bernardino, CA
03 March 2013